The Snake Test

Jimmy Huston

☐ True? ☐ False? ☐ Maybe

Cosworth Publishing
21545 Yucatan Avenue
Woodland Hills CA 91364
www.cosworthpublishing.com

For information regarding permission,
please send an email to *office@cosworthpublishing.com.*

☐ **True?** ☐ **False?** ☐ **Probably.**

Dedicated to Eve

(with all due apologies)

☐ **True?** ☐ **False?** ☐ **Who knows?**

So this is my story, told by me. And it wasn't easy.

Writing with a pencil is too hard for us snakes. Even if we get our mouth around it, we're just too close to the page to read.

So I had to type this book.

Not on a computer of course. I don't really understand computers.

On an old typewriter.

Also, I want you to know that all of the facts in this book are true.

It's the lies, exaggerations, and fabrications that are not.

☐ True? ☐ False? ☐ Both.

First of all, let me say this.

Leave me alone.

☐ True? ☐ False? ☐ Maybe.

2

Don't step on me. Don't grab me. Don't chase me.

Then we'll get along just fine.

☐ True? ☐ False? ☐ Certainly!

I'll bet the first thing you want to know is,
"Am I poisonous?"

☐ True? ☐ False? ☐ Ahm, yeah!

4

We'll talk about that later.

☐ True? ☐ False? ☐ Why not *NOW*?

5

You shouldn't always be so afraid of snakes.
We're actually pretty cool.

We're quiet. We don't bark, or moo, or cock-a-doodle-doo.

We stay out of your way as much as we can.

My skin is pretty. (I shed it occasionally.)

☐ True? ☐ False? ☐ Really?

6

We don't dig up your garden or chase your pets.

We don't smoke cigars or cigarettes.

Considering that we don't have legs, arms, hands, feet, ears, eyelids, wings, flippers, fins, fur, claws, or horns, we do pretty well.

☐ True? ☐ False? ☐ Dubious.

7

We can climb.

I can go up a tree or a fence post or a telephone pole.
That's good because I'm certainly not very tall.

☐ True? ☐ False? ☐ Neither.

8

Oddly enough, I can swim even though I can't paddle or kick.

Some snakes even live in the ocean. (But not at the beach.)

☐ True? ☐ False? ☐ I hope not!

9

I can't fly, of course, except down.

After I climb a tree I can fly straight down from it, onto the ground or into water.

That happens. It's a lot like falling.

☐ True? ☐ False? ☐ Oops.

There are some snakes called "flying snakes," but they don't really fly.

Flying snakes can only glide, cruising from one tree to a lower spot on another tree.

But it looks like flying. And it looks like fun.

☐ **True?** ☐ **False?** ☐ **Neither.**

I have a great sense of smell. In fact, right now -- as you're reading this -- I can smell you.

That seems to bother you a little, and I can sense your smell changing (not in a bad way).

Actually I'm not just smelling you. I'm tasting you.

If something smells interesting to me, then I flick my forked tongue. Why?

☐ True? ☐ False? ☐ Flick, flick, flick.

Because I have a super-smell sensor (called the Jacobson's organ) on the roof of my mouth.

So, when I stick out my tongue, it catches the smells in the air. When I bring it back into my mouth, my super-smeller goes to work and tells me what's out there.

Am I smelling dinner? A pizza perhaps? Or a delicious mouse?

☐ True? ☐ False? ☐ Sniff, sniff.

Why is my tongue forked?

Because it's super-cool looking?

Maybe, but also because -- depending on which side of the tongue a smell is on -- I can tell whether it comes from my left or my right.

And that's where dinner is.

I smell in stereo!

☐ True? ☐ False? ☐ Yuck!

We're great hunters. Snakes are patient and we're quiet.

Our jaws open so wide that our mouth is bigger than our head. We often swallow live prey. Sometimes we may only eat once a month.

We typically eat rodents, but will eat any small critter. Mice. Eggs. Bugs. Fish. Frogs. Grasshoppers. And birds.

(If a bird -- who can fly -- gets caught by a snake with no legs, maybe it deserves to be eaten.)

☐ True? ☐ False? ☐ Unlikely.

15

You may think a snake's tail starts at the head, but that is not true. The tail starts at the snake's hind legs.

☐ True? ☐ False? ☐ Legs?

16

Okay, we snakes don't actually have legs, but we used to have legs, long ago. Now our legs have reduced down to little "spurs," but they're there -- and that's where the tail starts.

☐ True? ☐ False? ☐ That's crazy!

There are over 3600 types of snakes.

We have a secret handshake, but none of us can do it.

☐ True? ☐ False? ☐ Yikes!

18

Sometimes, to preserve warmth, we hibernate in groups
called a nest, a den, a bed, a pit, or a slither. (Or, if
grouped for mating, a knot.)

☐ True? ☐ False? ☐ No thanks.

19

We're cold-blooded (called ectotherms or poikilotherms), so we like to lie in the sun.

And, we like to sleep -- a lot. Sometimes up to sixteen hours a day.

But we don't have eyelids, so we can't close our eyes, even to blink or wink. Luckily, we have transparent scales that protect our eyes.

☐ True? ☐ False? ☐ Both.

20

Snakes won't eat vegetables or other plants. We also won't eat at fast food restaurants or fancy hotels.

Sometimes we do eat other snakes. (Don't tell mom.)

Snakes never eat anything larger than giraffes. Maybe that's why you don't see many giraffes hanging around.

☐ True? ☐ False? ☐ Neither.

21

How can you tell the difference between a boy snake and a girl snake?

A boy snake's tail is longer and thicker.

☐ True? ☐ False? ☐ And his hat is bigger.

22

Sometimes, (not very often) a girl snake wears lipstick and high heels.

☐ **True?** ☐ **False?** ☐ *Ooh la la!*

23

Most types of snakes lay eggs that become "hatchlings."

Some types give birth to live snakes called "neolates." That sounds cute, doesn't it?

Baby snakes are "snakelets."

☐ True? ☐ False? ☐ Snakelets? I doubt it.

Snakes cannot ride a bicycle, but we're much better on a skateboard.

☐ True? ☐ False? ☐ So drop in.

25

The left lung of a snake can be smaller or completely missing, while the right lung does all the breathing. It runs the length of the snake.

That's why the good snakes never smoke.

☐ True? ☐ False? ☐ Absurd.

26

Snakes can't play the guitar or the piano in your band, but they can be pretty good drummers.

Especially the rattlesnakes.

☐ True? ☐ False? ☐ Errrrr...?

About those rattles.

Any snake with rattles is dangerous.

The rattles are there for a really good reason.

They tell you that the snake is annoyed. And that's a bad thing.

Leave that snake alone. Go away.

☐ True? ☐ False? ☐ Uh oh...

28

Sometimes rattlesnakes lose their rattles.

That makes them quiet.

But they can still bite.

And, the bite is still poisonous.

Watch where you step.

☐ True? ☐ False? ☐ Ouch!

I'll admit there are a few problems with being a snake.

We love burrows, but we can't dig.

We can't really chew our food.

We don't blink. And that means we can't wink. Or smile.

Or sing.

☐ True? ☐ False? ☐ Too bad.

30

I can't hear much, because I don't have ears.

(That also means I can't wear glasses because they keep falling off.)

I don't really need ears, because nobody ever talks to me anyway.

I do "feel" a lot of screaming when I see people running.

☐ **True?** ☐ **False?** ☐ **Huh?**

31

I have no hands, so I can't clap. No fingers, so I can't tie my shoes, which I don't have, because of no feet.

No feet means no dancing, but I can't hear music anyway. I can feel the vibrations -- like you might "hear" a passing car full of teenagers with a loud thumping bass.

☐ True? ☐ False? ☐ Both.

32

I can't do pushups.

I can't do chinups.

I can't even run. But, I might be as fast as you.

A fast snake can go 18 miles an hour.

☐ True? ☐ False? ☐ Can you run that fast?

33

Some snakes, called constrictors, coil around their prey and squeeze it to death before dining.

They love giant pretzels.

☐ True? ☐ False? ☐ Disturbing.

34

Snakes in some jungles can grow big enough to swallow a person whole.

If you live in a jungle, always check under your bed.

☐ True? ☐ False? ☐ Definitely!

Are we dangerous?

Maybe. It all depends.

How close are you?

What are you doing?

☐ True? ☐ False? ☐ I'm backing up.

Ask yourself this...

If someone was threatening you, and all you had to protect yourself with was teeth, what would you do?

You'd use your teeth. You'd bite.

And, if those teeth had poison in them...?

☐ True? ☐ False? ☐ Ouch!

37

That's the thing. All snakes can bite. (We sure can't hit or kick.)

But -- some snakes have two large front teeth (called fangs) that are hollow and contain poison (called venom) that can make a bite deadly. They're called vipers.

In North America there are four kinds.

☐ True? ☐ False? ☐ Too many.

When hunting, venom from a viper's bite kills its prey.

Luckily, when a viper swallows the prey, the poison doesn't harm the snake.

But -- if the snake accidentally bites itself, it can die from the poison, too.

☐ True? ☐ False? ☐ That's fair.

39

So -- after all that -- are you still wondering, "Am I poisonous?"

If you still don't know -- gently, slowly, close the book.

Quietly push it away.

And, above all, don't forget. Leave me alone.

THE END

☐ True? ☐ False? ☐ Adios.

P.S. I didn't really type this book on a typewriter.
That's just silly.

I dictated it.

☐ True? ☐ False? ☐ Whatever...

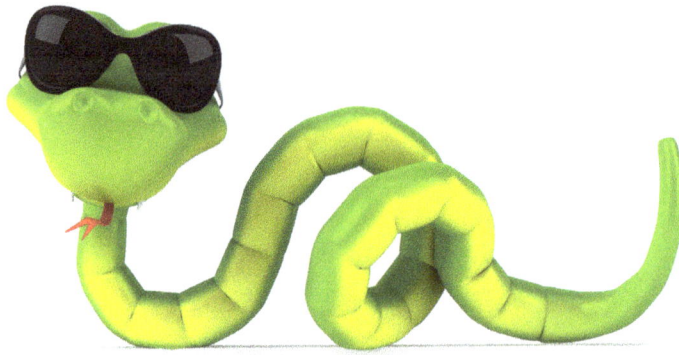

About the Author

This book was not written by a snake. Not really.

It was written by a guy who talks to snakes. And listens to them.

They told him what to write. He just does what he is told.

He lives in Woodland Hills, California, with his wife and dog and the cute little snake the dog keeps barking at in the hole under the bush in the back yard.

And he's written a bunch of other silly stuff. Check it out at *www.byjimmyhuston.com* or *www.cosworthpublishing.com*.

ANSWERS: Title Page - True; Info Page - True; Dedication Page - Who knows?; Page 1 Both; Page 2 True; Page 3 Certainly; Page 4 Ahm, yeah!; Page 5 Why not NOW?; Page 6 True; Page 7 True; Page 8 True; Page 9 True; Page 10 True; Page 11 True; Page 12 Flick, flick, flick; Page 13 True; Page 14 True; Page 15 True; Page 16 True; Page 17 True; Page 18 Yikes!; Page 19 True; Page 20 True; Page 21 True; Page 22 True; Page 23 Ooh la la!; Page 24 True; Page 25 So drop in; Page 26 True; Page 27 Errrr...?; Page 28 True; Page 29 True; Page 30 True; Page 31 True; Page 32 True; Page 33 True; Page 34 Disturbing; Page 35 Definitely!; Page 36 I'm backing up; Page 37 Ouch!; Page 38 True; Page 39 That's fair; Page 40 Adios; Page 41 Whatever...; Page 42 G'night.

☐ **True?** ☐ **False?** ☐ **G'night.**

More Books from Jimmy Huston

www.cosworthpublishing.com

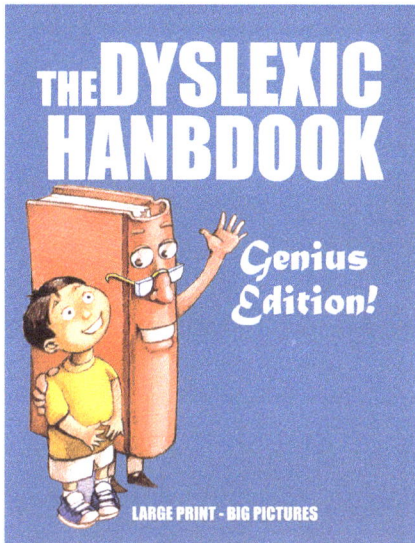

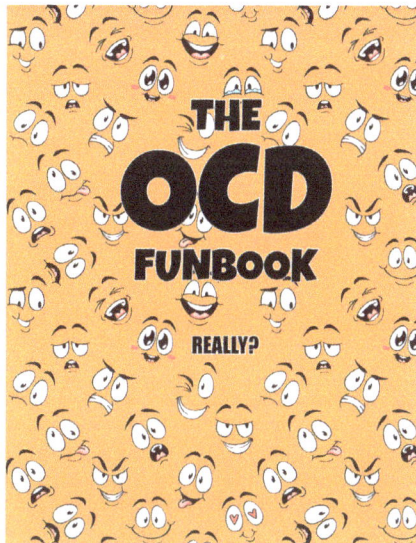

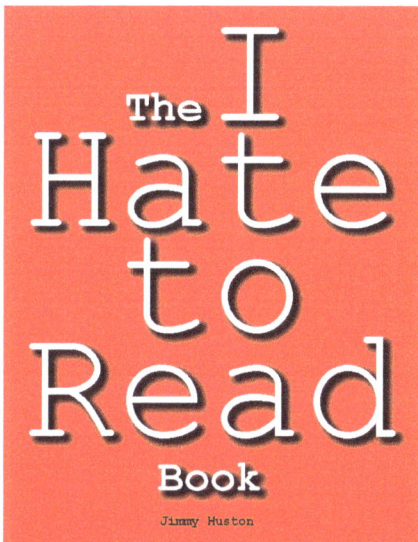

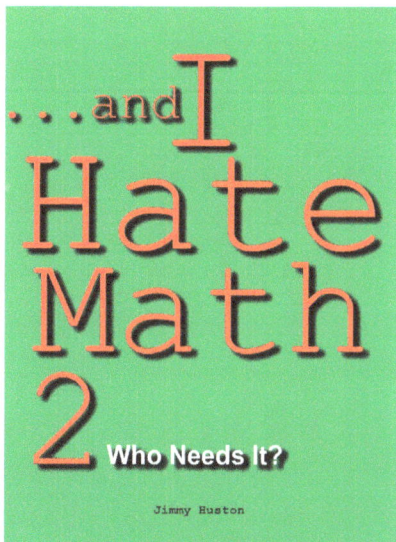

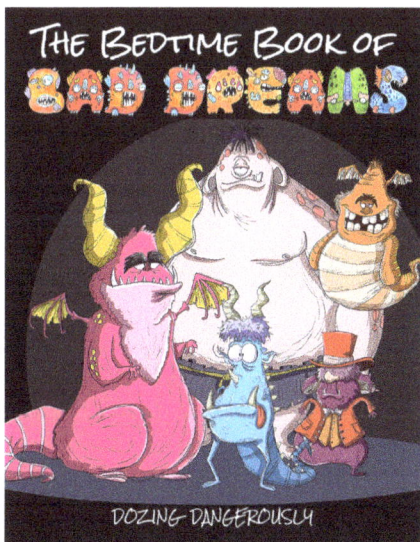

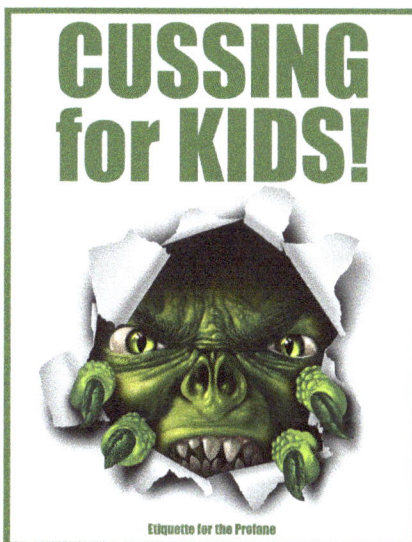

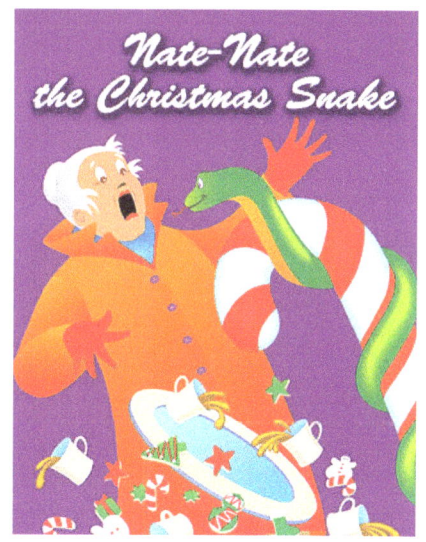

THE **DYSLEXIC HANBDOOK**
Genius Edition!
LARGE PRINT - BIG PICTURES

THE **OCD** FUNBOOK
REALLY?

The **Attention Deficit Disorder Hyperactive Cookbook**
Puzzle Edition

The **I Hate to Read** Book
Jimmy Huston

...and **I Hate Math 2**
Who Needs It?
Jimmy Huston

Rat BLEEP and Alien Poop
NOT FOR PARENTS AT ALL
A Non-Illustrated Picture Book
WARNING! The actual photographs of all Aliens have been **CENSORED**, so you will have to draw your own pictures.

THE BEDTIME BOOK OF **BAD DREAMS**
DOZING DANGEROUSLY

CUSSING for KIDS!
Etiquette for the Profane

Nate-Nate the Christmas Snake

More Books from Jimmy Huston

www.cosworthpublishing.com

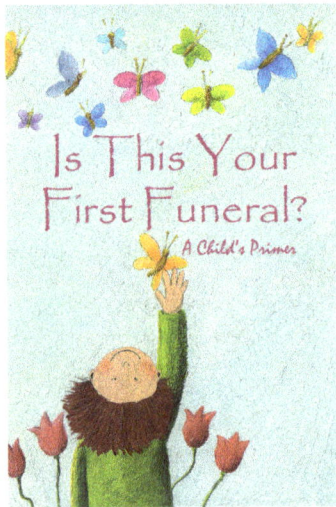

Is This Your First Funeral?
A Child's Primer

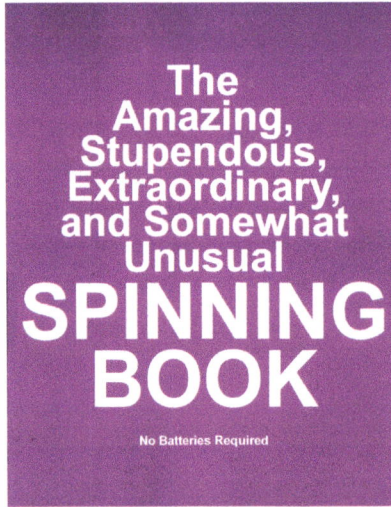

The Amazing, Stupendous, Extraordinary, and Somewhat Unusual SPINNING BOOK

No Batteries Required

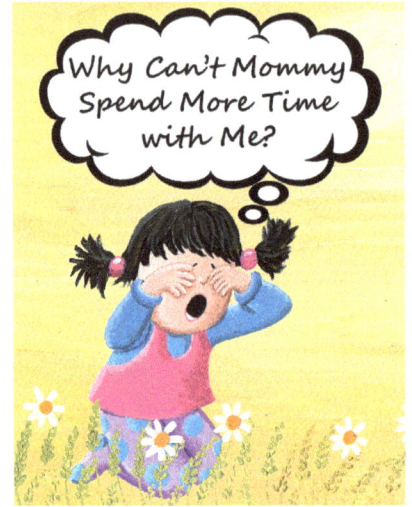

Why Can't Mommy Spend More Time with Me?

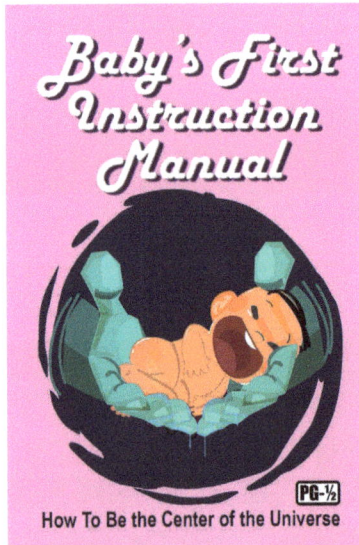

Baby's First Instruction Manual

PG-½

How To Be the Center of the Universe

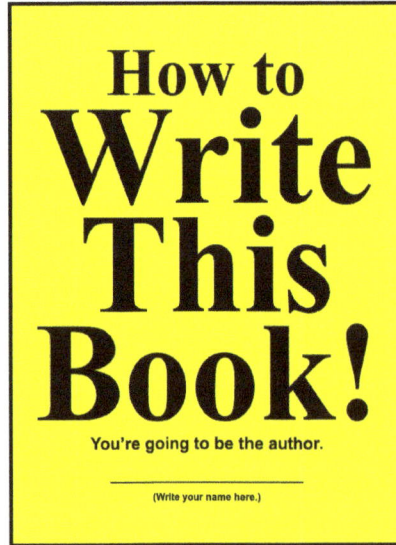

How to Write This Book!

You're going to be the author.

_____

(Write your name here.)

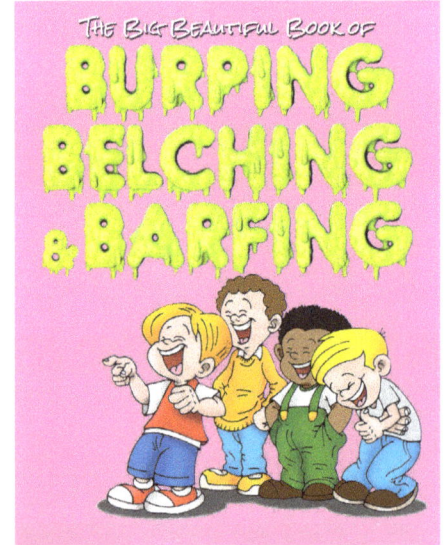

THE BIG BEAUTIFUL BOOK OF BURPING BELCHING & BARFING

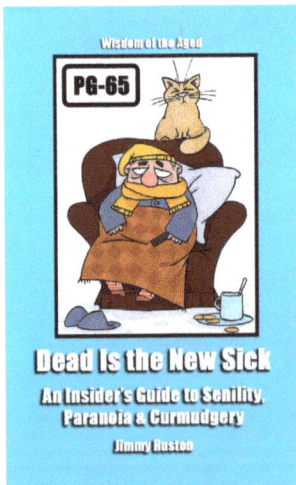

Wisdom of the Aged

PG-65

Dead Is the New Sick

An Insider's Guide to Senility, Paranoia & Curmudgery

Jimmy Huston

That Damn Little Angel!

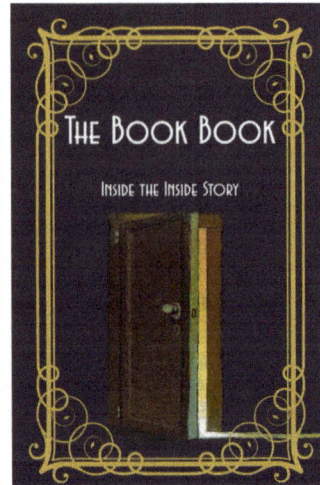

THE BOOK BOOK

INSIDE THE INSIDE STORY

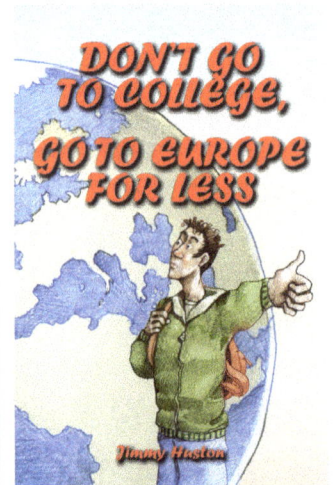

DON'T GO TO COLLEGE, GO TO EUROPE FOR LESS

Jimmy Huston

Books for Grownups from Cosworth Publishing

www.cosworthpublishing.com

THE SMART BRAIN PAIN SYNDROME
The Primer for Teens & Young Adults in Pain

· GEORGIA WESTON, LCSW · LONNIE K. ZELTZER, MD ·
· PAUL M. ZELTZER, MD ·

A groundbreaking new book. Three experts explain chronic pain to teens and parents, including using creative outlets to displace the pain.

PAIN: AN OWNER'S MANUAL
GEORGIA HUSTON

A young pain victim's inspirational and informative conversations with a variety of pain sufferers and specialists. Doctors should read this at their own risk.

Vienna's Waiting
A teenage girl's battle with pain.

georgia huston

At 14, Georgia developed mysterious chronic pain. This book chronicles that dark time and follows her inspirational journey back to health and happiness.

The Suicide Dilemma
Finding a Better Choice

Rebecca Morgan Gibson, LCSW
and
Lynn Mills

A practical guide for the person who is confronted by the possible suicide of a friend or family member.

Shanghai Torah
Yuanfen

Briana London

A young Jewish scribe flees WWII Europe with his in-progress Torah, escaping into China under Japanese occupation.

Vienna's Waiting
A teenage girl's battle with pain.

georgia huston weston
read by veronica huston

AUDIOBOOK

A powerful reading of Georgia's harrowing experiences as a young teen suffering chronic pain. Hearing it all out loud brings new power and meaning to this true-life story.

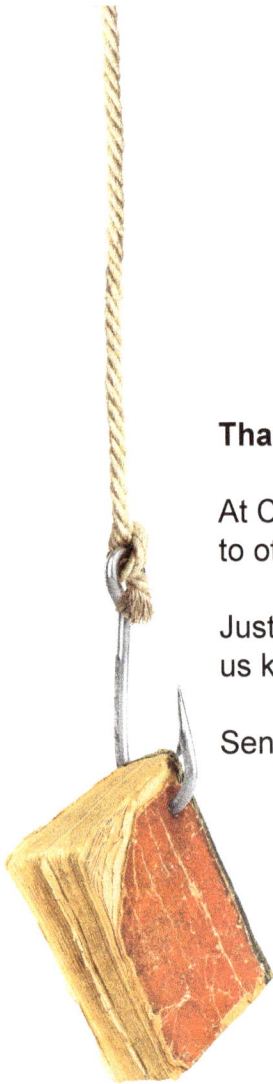

Thanks for buying, borrowing, or swiping this wonderful book.

At Cosworth Publishing we truly appreciate that, and in return, we'd like to offer you one of our E-books absolutely free—and worth every penny.

Just let us know that you want it, and we'll make sure that you get it. Let us know which book you read so we don't send you the same one.

Send an email to *office@cosworthpublishing.com*.

Then, from time to time, we will let you know via email when we have a new book that you might be interested in.

We won't do that very often because we're basically pretty lazy, and we don't produce very many new books.

Reviews are greatly appreciated.

www.ingramcontent.com/pod-product-compliance
Lightning Source LLC
Chambersburg PA
CBHW042334030426
42335CB00027B/3331